Nelson Science

This book belongs to:

_____

Workbook Starter B

Anthony Russell

OXFORD
UNIVERSITY PRESS

OXFORD
UNIVERSITY PRESS

Great Clarendon Street, Oxford, OX2 6DP, United Kingdom

Oxford University Press is a department of the University of Oxford.

It furthers the University's objective of excellence in research, scholarship, and education by publishing worldwide. Oxford is a registered trade mark of Oxford University Press in the UK and in certain other countries.

British Library Cataloguing in Publication Data

Data available

ISBN: 978-1-3820-1778-7
ISBN: 978-1-3820-1779-4 (Workbook only)

1 3 5 7 9 10 8 6 4 2

Paper used in the production of this book is a natural, recyclable product made from wood grown in sustainable forests. The manufacturing process conforms to the environmental regulations of the country of origin.

Printed in Great Britain by Bell and Bain, Ltd Glasgow

Acknowledgements

The publisher and authors would like to thank the following for permission to use photographs and other copyright material:

Cover: Aaron Cushley. **Photos: p3(tl):** mama_mia/Shutterstock; **p3(tr):** New Africa/ Shutterstock; **p3(bl):** Joe Hendrickson/Shutterstock; **p3(br):** Artazum/Shutterstock; **p11:** Esther Moreno / Alamy Stock Photo; **p30(tl):** Maxena/Shutterstock; **p30(tr):** Sandymsj/Shutterstock; **p30(ml):** Vomirak/Shutterstock; **p30(mr):** Bayliss photography/Shutterstock; **p30(b):** Romrodphoto/Shutterstock; **p32:** Robert_s/Shutterstock; **p43(t):** Jiri Hera/Shutterstock; **p43(b):** Tongra Jantaduang/Shutterstock.

Artwork by Q2A Media Services Pvt. Ltd. and David Semple.

Every effort has been made to contact copyright holders of material reproduced in this book. Any omissions will be rectified in subsequent printings if notice is given to the publisher.

MIX
Paper from
responsible sources
FSC
www.fsc.org   FSC® C007785

Contents

This is where I live

1 Draw your home.

2 Colour it.

3 Talk about your drawing with a partner.

Hard things and soft things

1 Name what you see in the photos.

2 Put ☒ on one hard thing.

3 Draw a circle round one soft thing.

1 Draw one thing you have in each room at home.

kitchen	bedroom
bathroom	living room

2 Find someone who has drawn a different thing in one of the rooms. Talk about your drawings.

In and out, up and down

1 Tick the things you have at home.

2 Put ☒ on the things that help us go up from the ground.

3 Draw your favourite thing about where you live.
 Show and tell.

This is my school

1 Draw people at school.

1 How many children in each row?

2 Write the numbers. Count from the front row.

_____ _____ _____

3 Complete the drawings.

How many girls? _____

How many boys? _____

1 Talk to a partner about what you see in the picture.

2 Draw a circle round the activity you like most.

3 Draw children playing with three different things.

4 Colour the drawing. Show and tell.

Books and pictures

1 Following the steps shown here, make a book.

a) Fold some pieces of paper.

b) Arrange them like this.

c) Ask an adult to staple the pages together.

d) Give the book a title.

e) Draw pictures inside.

2 Show and tell.

Reading books

1 Work with a partner. What kind of books does your partner like?

2 Where do you like to read?

Getting information

1 A dictionary is a type of book which gives us information.

2 Name the animals you can see.

3 Draw an animal beginning with the letter **a**. Show and tell.

1 Draw a picture of a fruit beginning with **a** and **b** on each page.

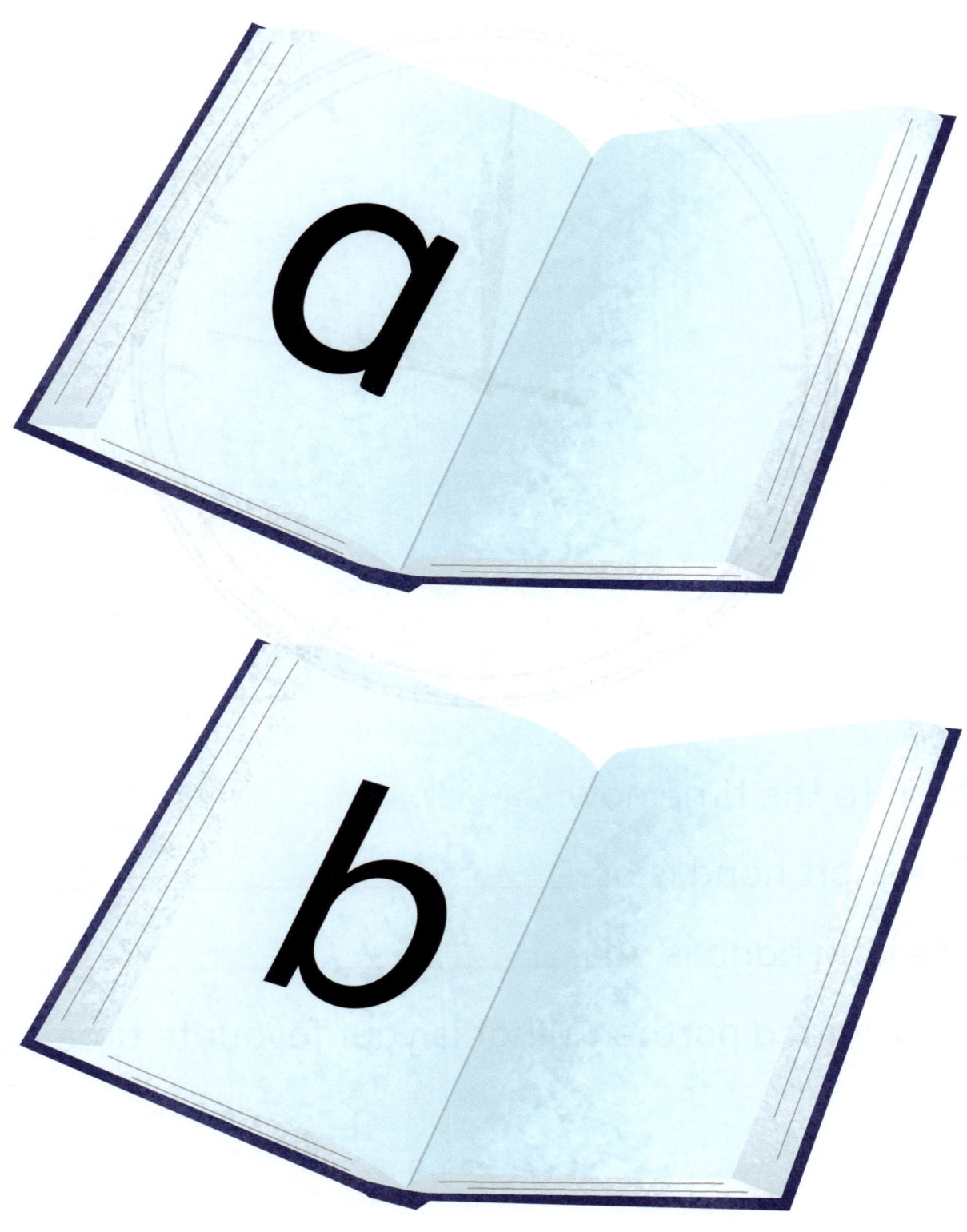

2 Show and tell.

Clocks

1 Put the numbers on the clock.

2 Point to the time now.

The short hand is on _____ .

The long hand is on _____ .

3 Work with a partner. What is your favourite time?

1 What time do you do these things?

2 Put the times in the boxes.

3 Draw the hands on the clock. You choose the time.

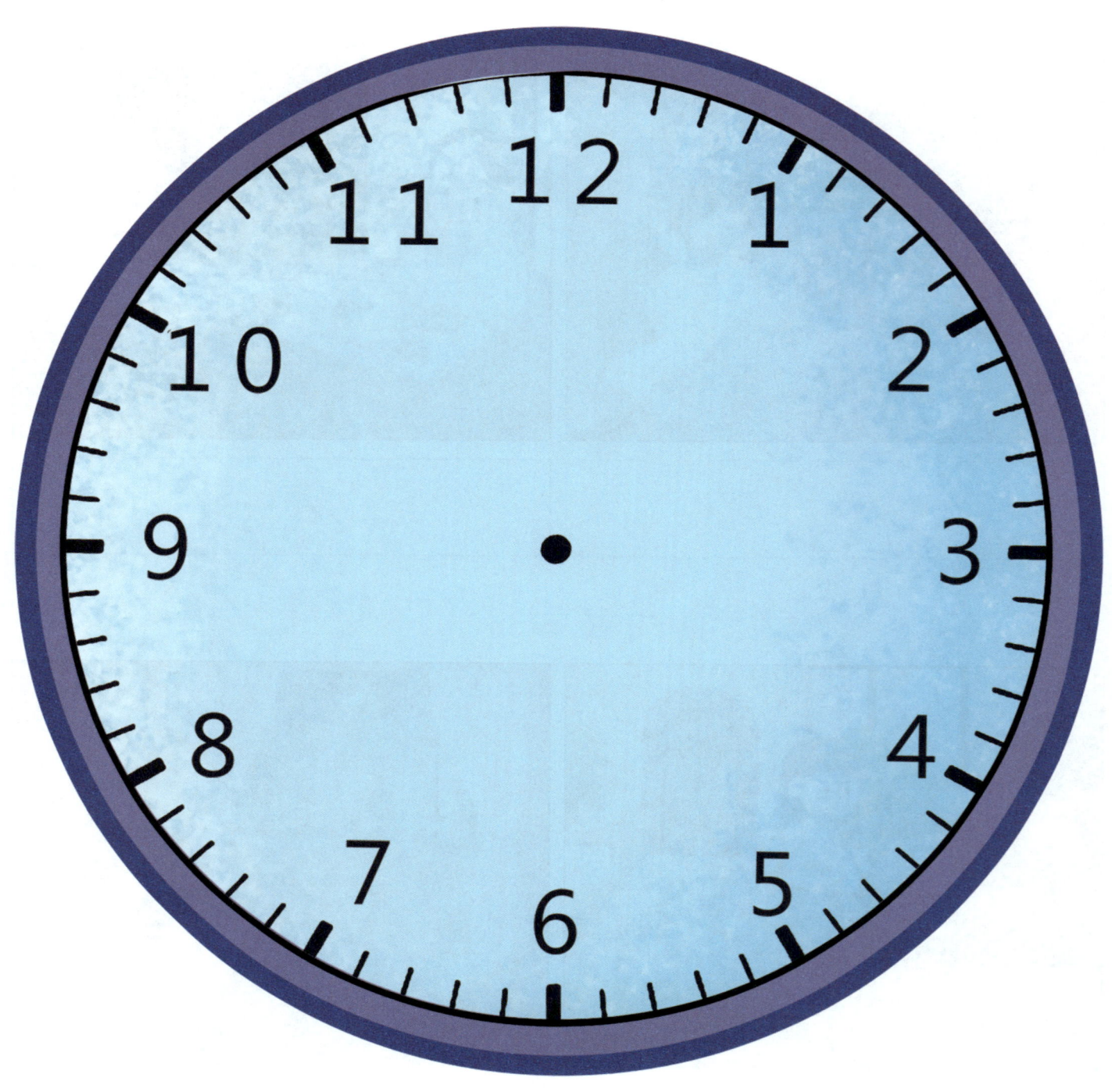

4 Work with a partner.
 Show and tell.

5 Ask and answer. What time is it?

UNIT 5

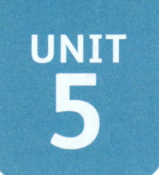

Time for food

1 Name the foods you can see.

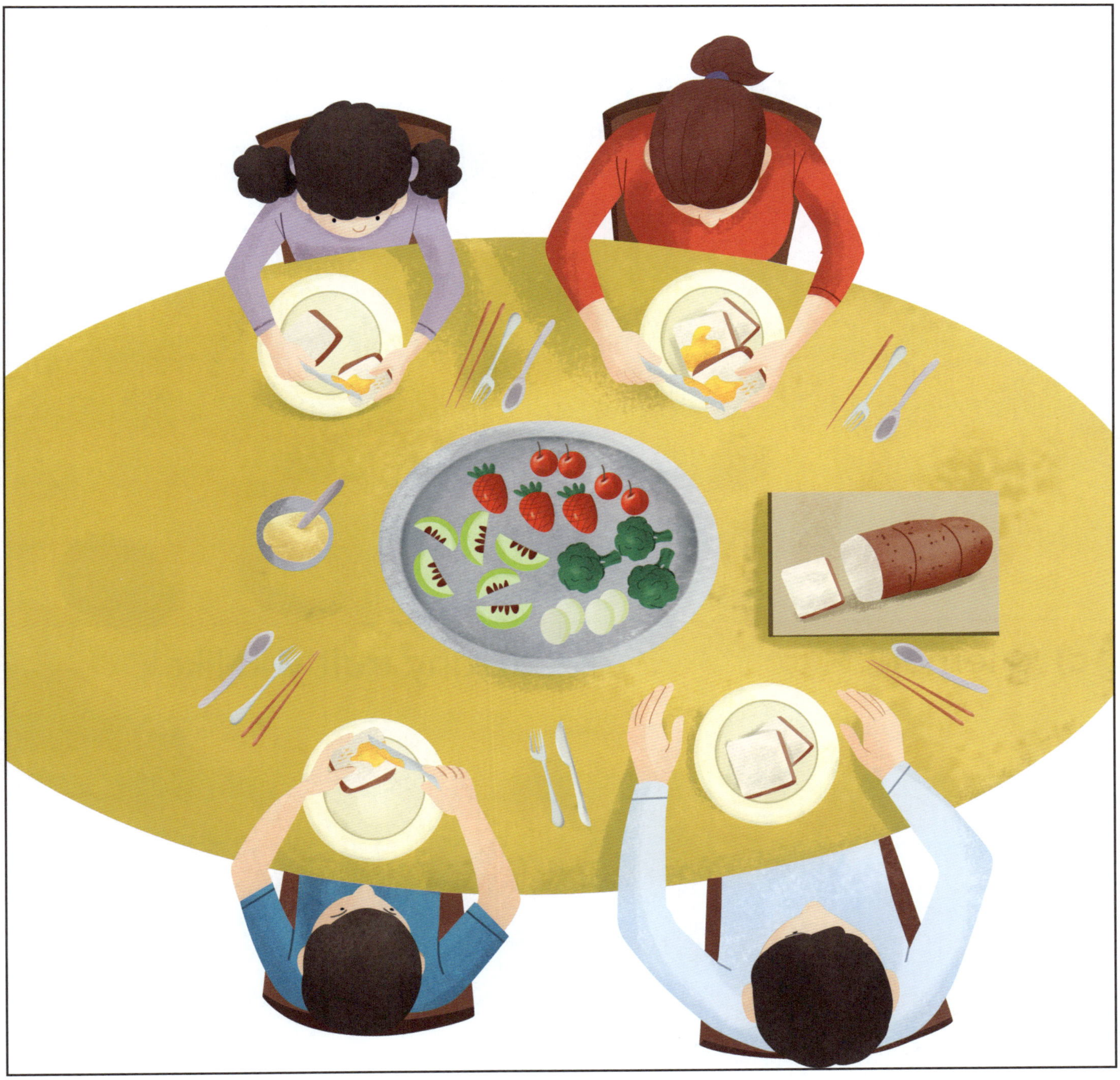

2 Put ☒ on foods from plants.

1 Draw and colour each fruit in the boxes below.

a) a banana	**b)** an orange
c) a lemon	**d)** any fruit you like

2 📖 Show and tell.

1 Count each type of thing.

2 Write the number of each set.

_____ spoons

_____ knives

_____ plates

_____ bowls

_____ tumblers

_____ chopsticks

3 Draw one piece of green food, one piece of red food and one piece of yellow food on the table above.

4 Show and tell.

1 Name things you can see.

2 Put ☒ on the dangers in the picture.

3 Draw your favourite food. 👤 Show and tell.

Time for school

1 Fill in the missing letters.

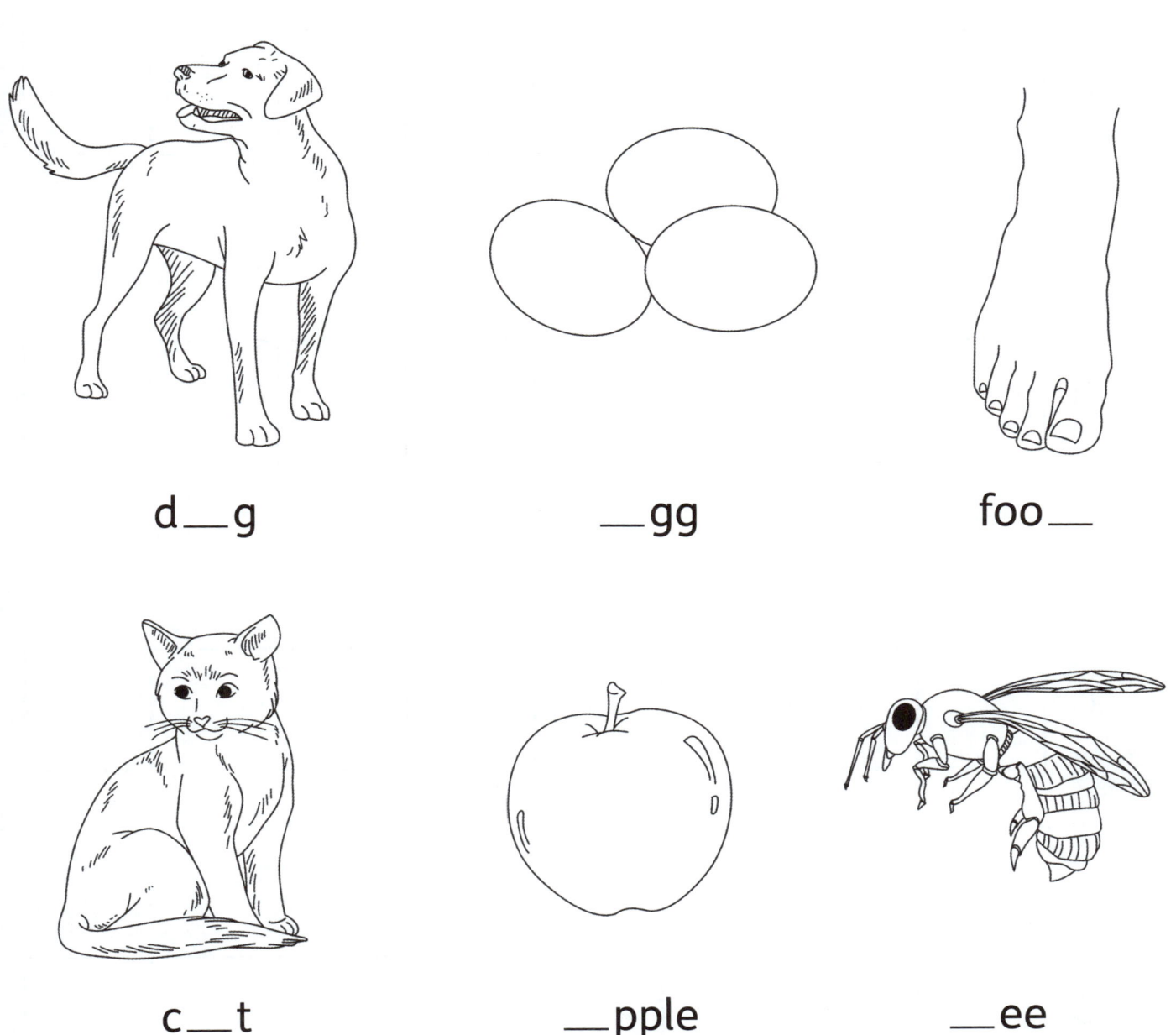

d __ g

__ gg

foo __

c __ t

__ pple

__ ee

2 Colour the pictures.

Measuring

1 Measure these things using the squares:

a) hand   **b)** crayon   **c)** leaf   **d)** you choose something.

A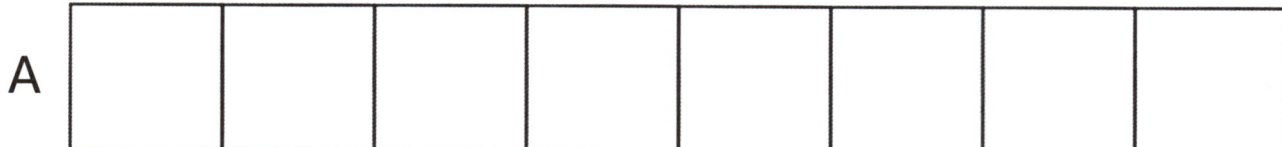

B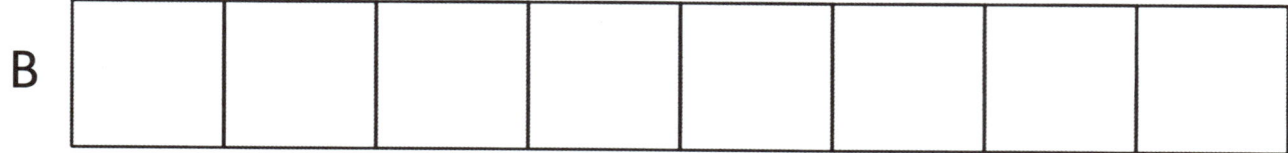

C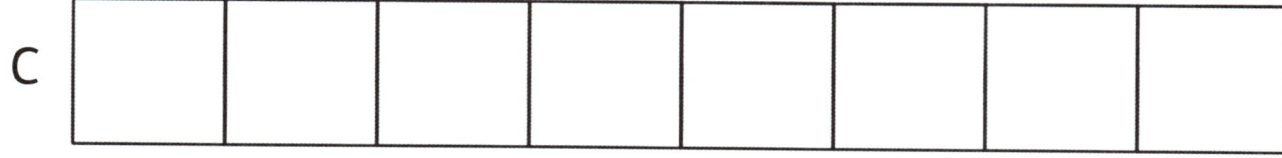

D

2 Colour the squares for each thing.

3 Count the squares.

4 Which is the longest? Mark it with ☒.

Shape sets

1 Number each set of shapes in order of size. Number them from 1 to 4, starting with the smallest. The ovals have been done for you.

2 Colour each set a different colour.

1 Count the leaves.

2 Count the flowers.

3 Write the numbers.

| _____ leaves | _____ flowers |

4 Count the sets of each colour.

5 Write the numbers.

_____ blue flowers	_____ orange flowers
_____ white flowers	_____ red leaves
_____ green leaves	_____ yellow leaves

Playtime

1 Follow these steps to make a windmill.

You will need: pencil   ruler   split pin   scissors
square piece of card

2 Colour it before you pin it.

3 Go outside and run with your new windmill. What happens when you run with it?

Shapes from blocks

1 Make 3 different shapes from 6 blocks.

2 Draw one of the shapes from blocks. Colour it.

3 Find someone who has a different shape from you.

1 Name the activities in the picture.

2 Work with a partner. Talk about which two activities you most enjoy.

3 Draw your favourite activity.

1 Draw a made-up animal.

2 Colour it.

3 Tick the ways your animal moves:
walk   swim   fly   slide   hop   run

4 Show and tell.

UNIT 8

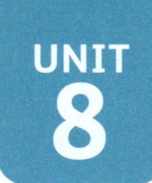

Time for sleep

1 Name the things you see.

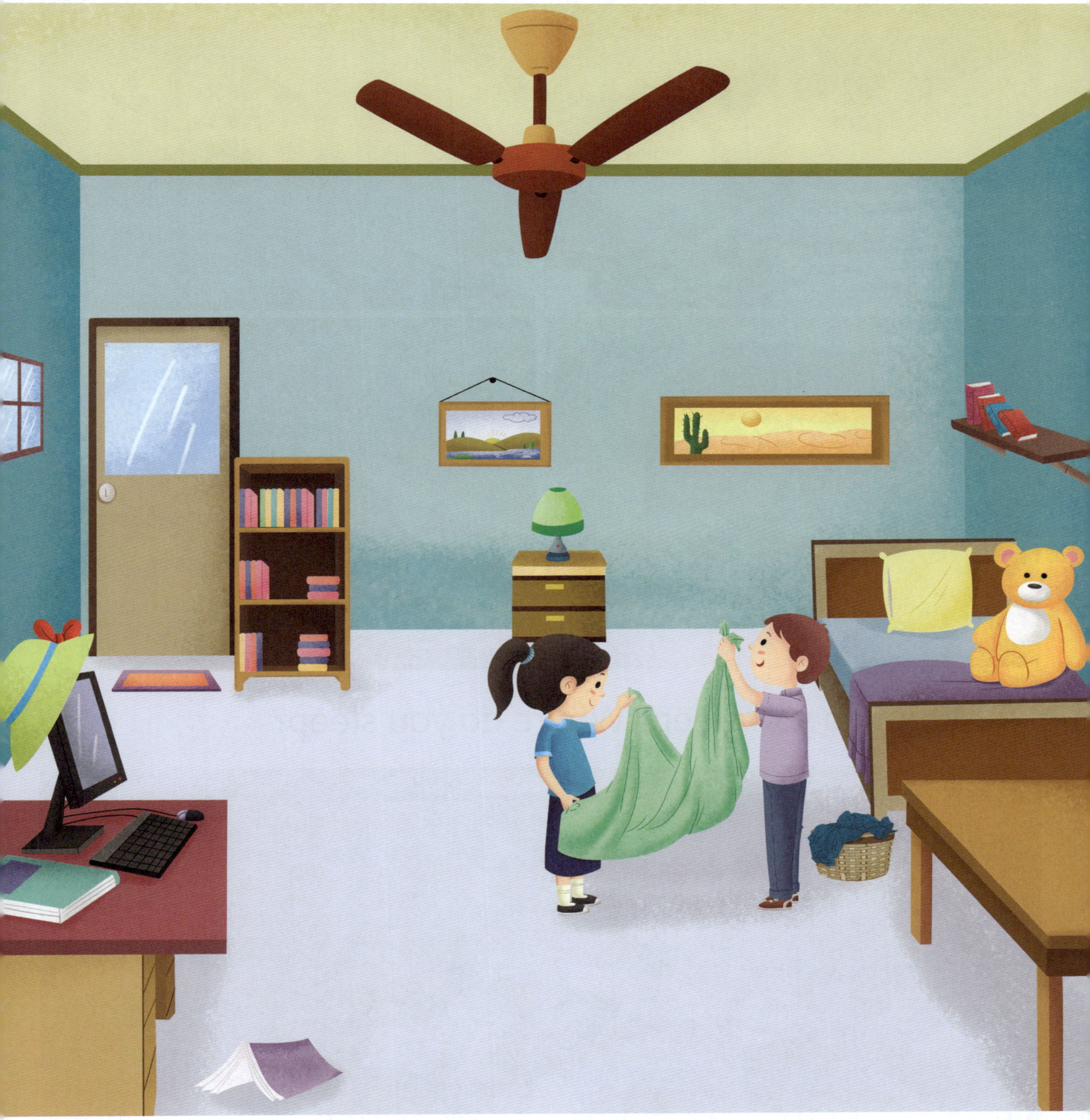

2 Put ☒ on the hard things. Circle the soft things.

Sleeping places

1 Put a ☒ by any animal you have seen.

2 When do you sleep? Where do you sleep?

1 Cross out the things which have nothing to do with sleep time.

2 Colour the sleep items.

3 Looking at the Moon and stars helps us rest. What helps you go to sleep? What wakes you up?

4 Draw a picture of a dream.

5 Show and tell.

UNIT 9

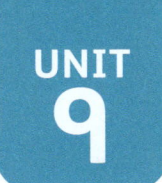

Special days

1 Cross out all the things you do **not** see on the table at a party.

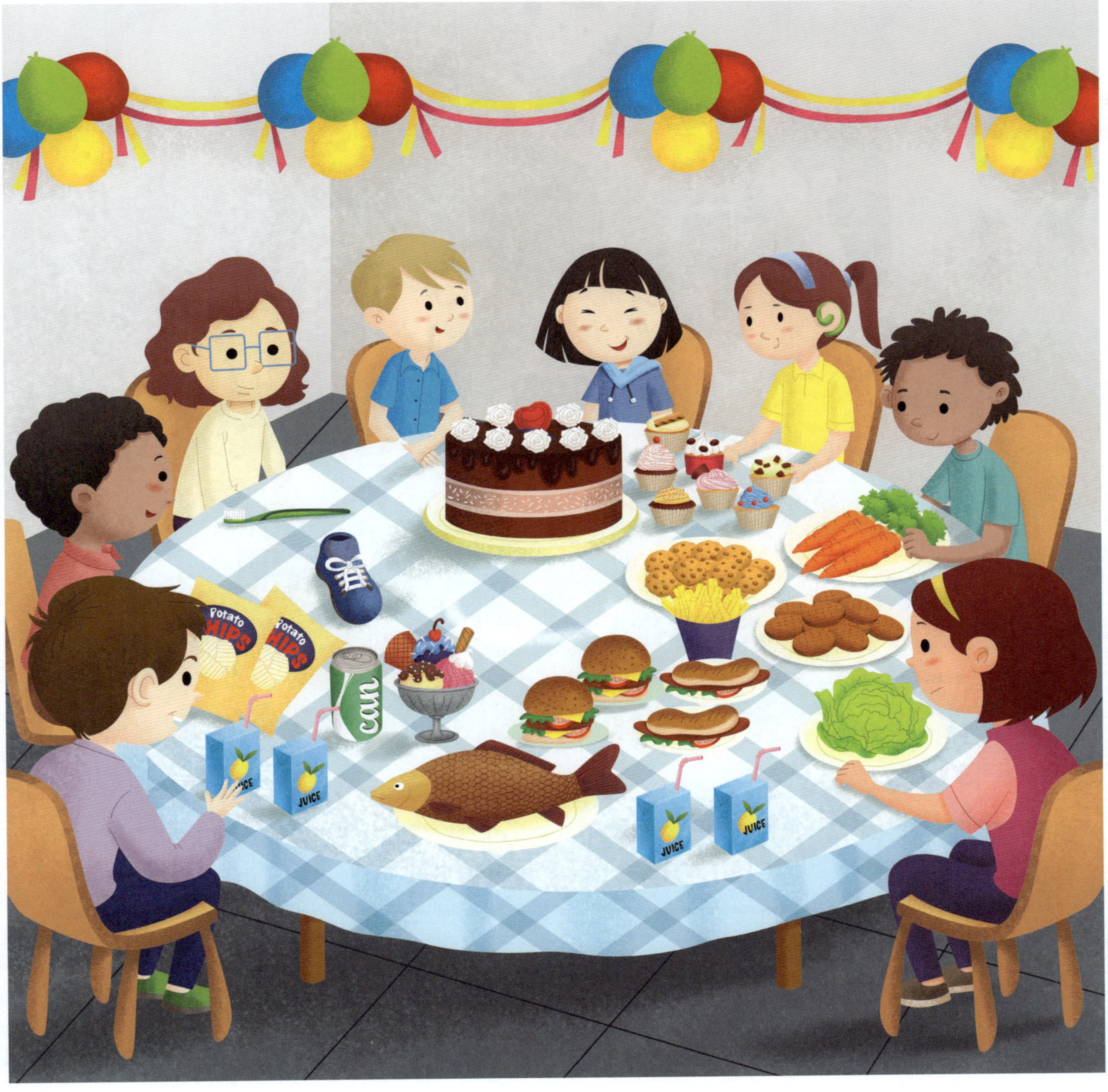

2 Put ☑ on the thing you like best. Put ☒ on the thing you would not like at your party.

A wedding

1 What is happening in this picture? Talk about it with a partner.

| playing | clapping | singing | dancing |

2 Draw lines from the words in the boxes to the picture.

1 What would you like to do at the fair? Circle your favourite thing.

round and round

up and down

up and down and round and round

2 Draw lines from the words in the boxes to the picture.

1 Mark the moving things with a ☒.

2 Look for the pattern in the little flags. Continue the pattern.

3 Draw a costume you would wear to a carnival. Show and tell.

UNIT 10

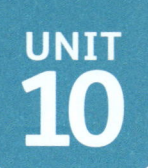

Growing

1 Look at the pictures. Put A by the pictures of people. Put B by the pictures that are **not** people.

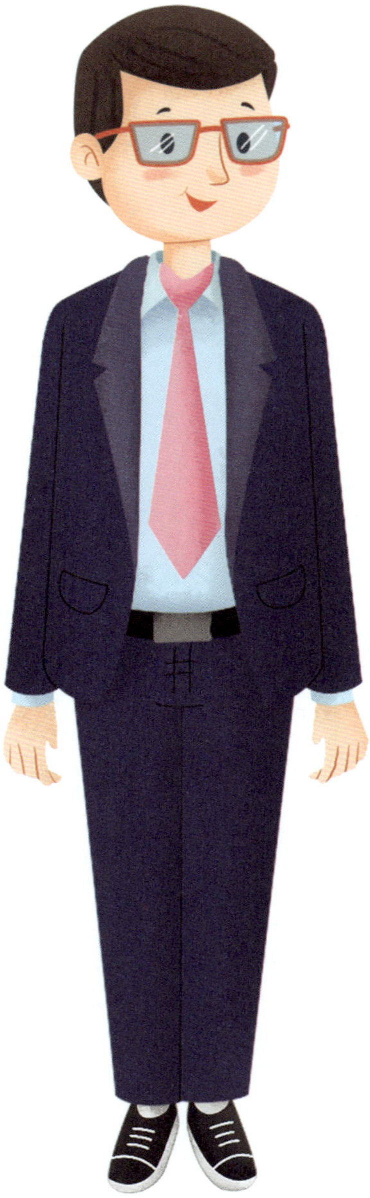

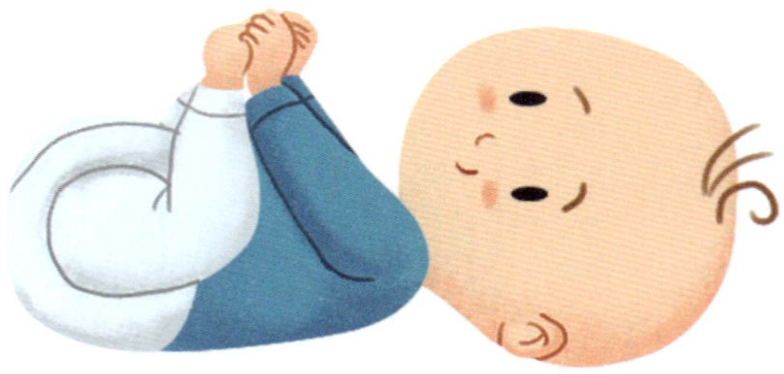

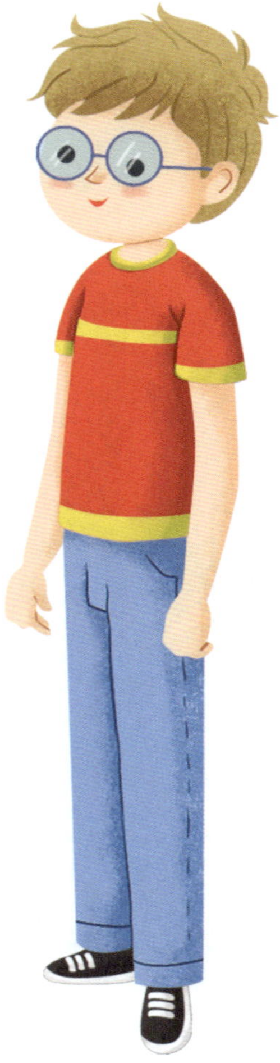

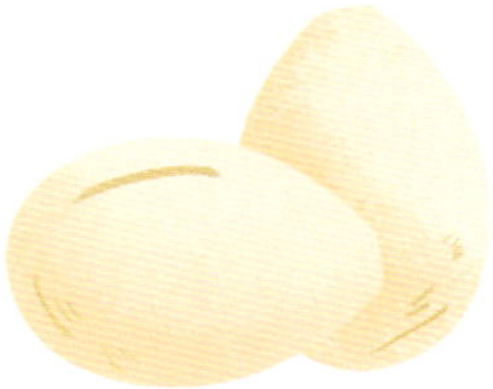

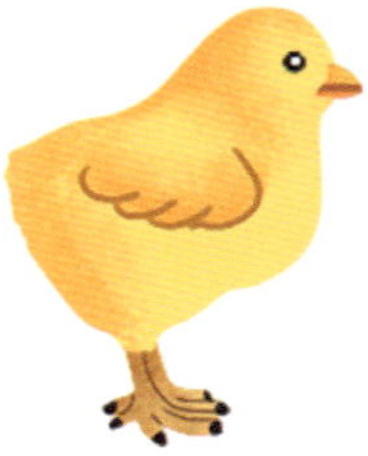

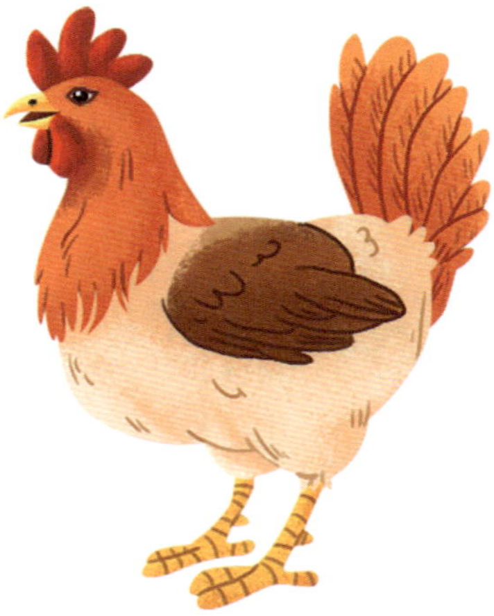

2 Put each set in order of age. Write (1) (2) or (3) beside them. 1 should be the youngest, 3 the oldest.

Height

1 Put the family in order of height.

2 Write 1–5 beside each family member. 1 is the shortest, 5 is the tallest.

3 Draw yourself and a family member who is taller than you.

Wheels

1 How many wheels can you see on this page? Point and add.

2 Work with a partner. Talk about which picture you would like to ride on.

3 Colour the pictures.

Seeds

1 Look at the pictures showing a seed being planted.

2 Next the plant grows taller. Draw this.

3 👤 Show and tell.

Changes

1 Tell the life story of the bird.

2 Which is the adult bird from our story? Tick the correct bird.

☐ ☐ ☐

3 Draw some eggs in a nest. Colour them.

Growing and changing

1 Look at the story below. Number the steps of the story from 1 to 4.

The story of a plant.

The plant grows flowers.

I put the seed in the soil.

The plant grows leaves.

The seed must have water.

Push and pull

1 Write the correct word: push or pull.

_____

_____

_____

_____

_____

_____

2 Find some things in the classroom. Try pulling them. Draw one of them.

3 Find some things in the classroom. Try pushing them. Draw one of them.

Glossary

adult – adults are fully grown animals and people

animals – living things that must eat plants or other animals as food

birds – animals which have feathers and lay eggs on land

egg – the starting point for a new animal

flower – the part of the plant that can produce a fruit

food – animals and plants need food to help them grow.

fruit – the part of the plant that has seeds in it

grow – living things grow and get larger

hard – cannot easily be cut or squashed

leaf/leaves – plants make food in their leaves

plants – green living things that make food using sunlight, water and a gas from the air

pull – a force that moves an object *towards* the source of the force

push – a force that moves an object *away from* the source of the force

seeds – the parts of a plant made in fruits and which can grow into new plants

soft – can easily be squashed, bent or shaped

stars – stars are suns that are so far away they look like tiny points of light in the night sky

water – a liquid which is essential for all living things